I0118657

Before Cancer

Dr. Kevin A. Figueiredo, PhD

Copyright 2016   Kevin A. Figueiredo

All Rights Reserved. No part of this publication may be reproduced, distributed, or transmitted in any form or by any means, including photocopying, recording, or other electronic or mechanical methods, without the prior written permission of the publisher, except in the case of brief quotations embodied in critical reviews and certain other noncommercial uses permitted by copyright law. For permission requests, write to the author at the address below.

Dr. Kevin A. Figueiredo
720 King St. W, Suite 416,
Toronto, ON
Canada M5V 3S5

Terminate Cancer Series: Volume 2

First Edition
Published Digitally – April 2016
Published in Print – April 2016

FTH Publishing
www.doctor-kevin.com

ISBN-10: 0-9975574-0-0
ISBN-13: 978-0-9975574-0-4

This book is dedicated to all those who recognize

the resilience of the human body and spirit

regardless of the presence or absence of disease.

CONTENTS

Chapter 1

Health Science Introduction

Health is the absence of disease. Science is the study of universe which includes human life. Therefore, ***health science*** is the study of human life in the absence of disease. Health science begins with genetics and environmental effects like nutrition. As an individual grows, additional environmental factors also become a critical part of health science including exercise and air quality, as well as absence of cigarette smoke, alcohol, drugs and stress. This book aims to assist readers with achieving healthy lifestyles based on the foundation of health science. Therefore, it is important that we understand the above definition of health science before beginning our discussion on healthy living. It is the definition we will be using for describing health science throughout this book.

The blueprint for life is encoded in genes which make up our DNA after fertilization. Expression of these genes results in production of proteins and subcellular structures which create the individual cells of the body.

These cells in turn eventually form tissue layers during embryonic development resulting in the individual. The requirements for health remains much the same as the individual progresses from childhood to adult. This includes quality nutrition, good air quality and exercise.

Some cancers such as early childhood cancers may be unavoidable due to genetic inheritance, however the vast majority of cancers can be prevented even in individuals that are predisposed to them. The critical factor are those everyday lifestyle choices that we all make (figure 1). For those that are carriers of genes that can be potentially cancerous later in life, it is vital not to make yourself susceptible to poor lifestyle choices because it is the latter that will eventually trigger the cancer. Our primary objective must be to exercise absolute control over the food quality and exercise regimen that we each select for ourselves.

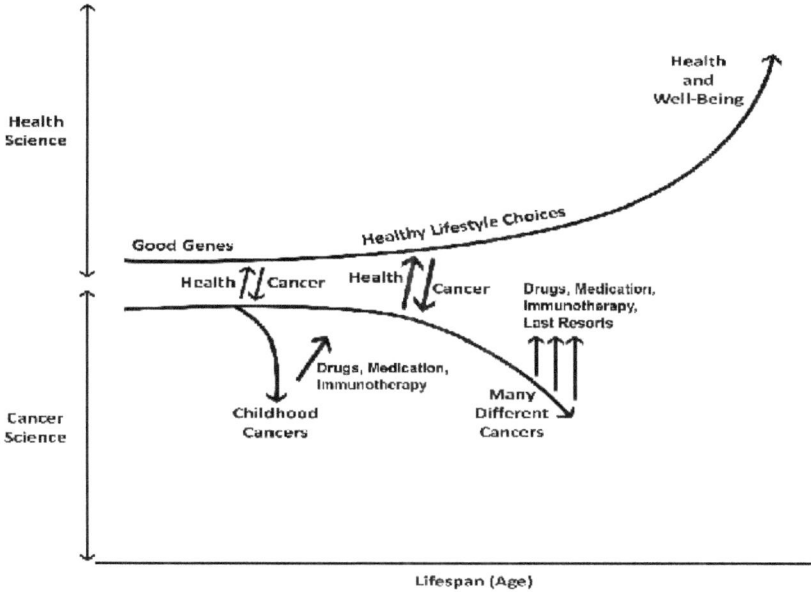

Figure 1 – The Health/Cancer Equilibriums. The top half of the graph depicts the benefits of healthy lifestyle choices, and the bottom half depicts mostly the consequences of poor lifestyle choices. Throughout life there is always an equilibrium between health and diseases such as cancer. It is our everyday decisions and habits that will determine whether each of our molecular equilibriums will shift towards health or disease at any time period throughout our lives. This really underscores the importance of making the correct lifestyle choice, not just today or this week, but for everyday throughout your lifetime.

In the past, it was believed that most of us cannot control our genetics, however recent studies reveal that this may now in fact be a possibility and we'll discuss this in detail later in this book. This process known as epigenetics is likely also a result of the nutrition we consume, and this gives us a new second layer of environmental control. So in the twenty first century, we don't have much excuse for not exercising our nutritional controls. Nutrition ultimately regulates our cellular composition and determines the health of our individual cells through modulation of the cell cycle (figure 2). Throughout life, there are millions of cells in our bodies that are consistently undergoing turnover from a normal healthy state, a process of cellular division (known as mitosis), and eventually cell death (known as apoptosis) where the older cells are replaced with a new generation of cells. This cycle of life at the cellular level is known as the cell cycle (figure 2). So the phrase 'you are what you eat' is not just a cliché, it is a literal fact that is increasingly being proven true by modern science.

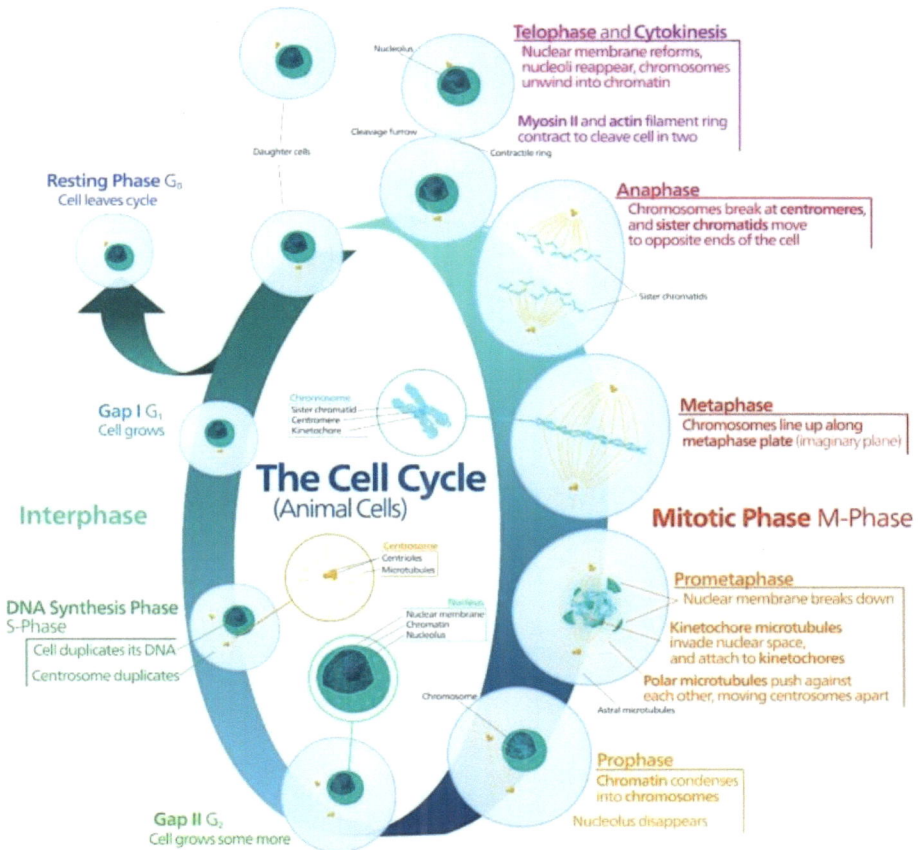

Figure 2 – The Cell Cycle. The mitotic phase is a relatively brief period of the cell cycle. During interphase, which includes G1 (first gap), S (synthesis) and G2 (second gap) the cell grows through production of proteins and organelles. Chromosomes are replicated only during the S phase. After the phases of growth, synthesis, and more growth during interphase, the cell is then ready for division during mitosis, and then restarts the cycle once again. Occasionally, cells will leave the cell cycle and enter Go phase. Entry into Go occurs in normal tissue through a mechanism of density dependent inhibition. When multiple mutations occur in a given cell it can become cancerous. When the regulatory process of the cell cycle malfunctions in cancer cells, they may continue through the cell cycle without concern for the density of surrounding cells, thereby possibly creating a tumour.

During an individual's lifetime, every now and then, one or two cells will resist the destruction pathway and may start to become cancerous. These are often benign and can easily be brought back to a normal cell cycle, if the individual has proper nutrition and a healthy lifestyle. If the person makes poor lifestyle choices throughout life, then this health/cancer equilibrium may shift towards the cancerous state (figure 1). In individuals with poor lifestyle choices, cancer cells can thrive due to presence of free radicals in the blood, and poorly oxygenated, acidic and inflammatory microenvironments which are ideal for tumour growth and proliferation. If the individual continues to make poor lifestyle choices, then eventually they will be led to a path which may require some conventional therapeutic options, such as drugs, chemotherapy, and radiotherapy.

In my previous book, *Terminate Cancer*, we discussed a new model of viral infection and immune response as a potential means to treat cancer. This immune system regulated therapy (immunotherapy) is an alternative for cancer patients that may spare them some of the side effects of conventional therapies.

Nonetheless, all these last resort therapeutic options may not be required in the first place, if we only paid closer attention to the daily lifestyle choices we make earlier in life when we are healthy (figure 1).

Billions of funds each year are spent on cancer science – the far edge of health. Unfortunately, this is very twentieth century thinking. Going forward in the new millennium, it will be very important for the public to lobby for funding for health promotion, instead of solely focusing funds on disease. Similarly, money can help you buy pharmaceutical drugs when you are ill, or it can buy you nutritious food when you are healthy. The bottom line is that it would be great if more government spending was allocated toward health promotion, however in the end we each have to take responsibility for our own health. The more we focus on health, the less likely we will have to deal with disease. Ultimately, this is a choice each individual will have to determine for themselves and their family… preferably *Before Cancer.*

Dr. Kevin A. Figueiredo, PhD

Chapter 2

Genetics and Epigenetics

Genetics is the study of genes or DNA, including trait inheritance and molecular inheritance mechanisms. Modern genetics has also expanded beyond inheritance to studying the function and behavior of genes.

Many childhood cancers originate from misfortunate genes. Within a cell, genes can be switched on or off by regulatory factors. If there is overproduction of a regulatory factor or if there is a mutation in the regulatory element on the DNA where it binds, then this can often result in expression of oncogenes or it can impair the expression of tumour suppressor genes. Either of these scenarios could potentially lead to cancer. Oncogene production results from a gain of function mutation. Most normal cells would undergo the process of programmed cell death known as apoptosis, however cells with oncogenes may avoid this pathway and proliferate instead. Tumour suppressor genes are a result of loss of function mutations where loss of two copies of the gene (alleles) are required for cancer to occur.

The proteins encoded by tumour suppressor genes usually have a suppressing effect on proteins involved in cell cycle progression, and also promote apoptosis.

Most cancers in adults are triggered by environmental factors like nutrition. Genetics often plays a critical role in predisposition and susceptibility of the individual to such environmental factors. The prevailing thought for many decades is that we have no control over our genes. However, new research suggests that epigenetics plays a crucial part in determining the final destiny of our cells. Epigenetics blurs the line between genetics and environmental factors (figure 3). As with classic and modern genetics, the net impact of epigenetics is on DNA transcription and protein expression. So in effect, epigenetics acts as a responsive interface between our DNA and environmental factors like the foods that we consume. Epigenetic mechanisms are affected by several factors such as environmental chemicals, pharmaceuticals, aging, and nutrition. Methyl groups are an epigenetic factor found in some dietary sources that can tag DNA and activate or repress genes. This process is known as DNA methylation.

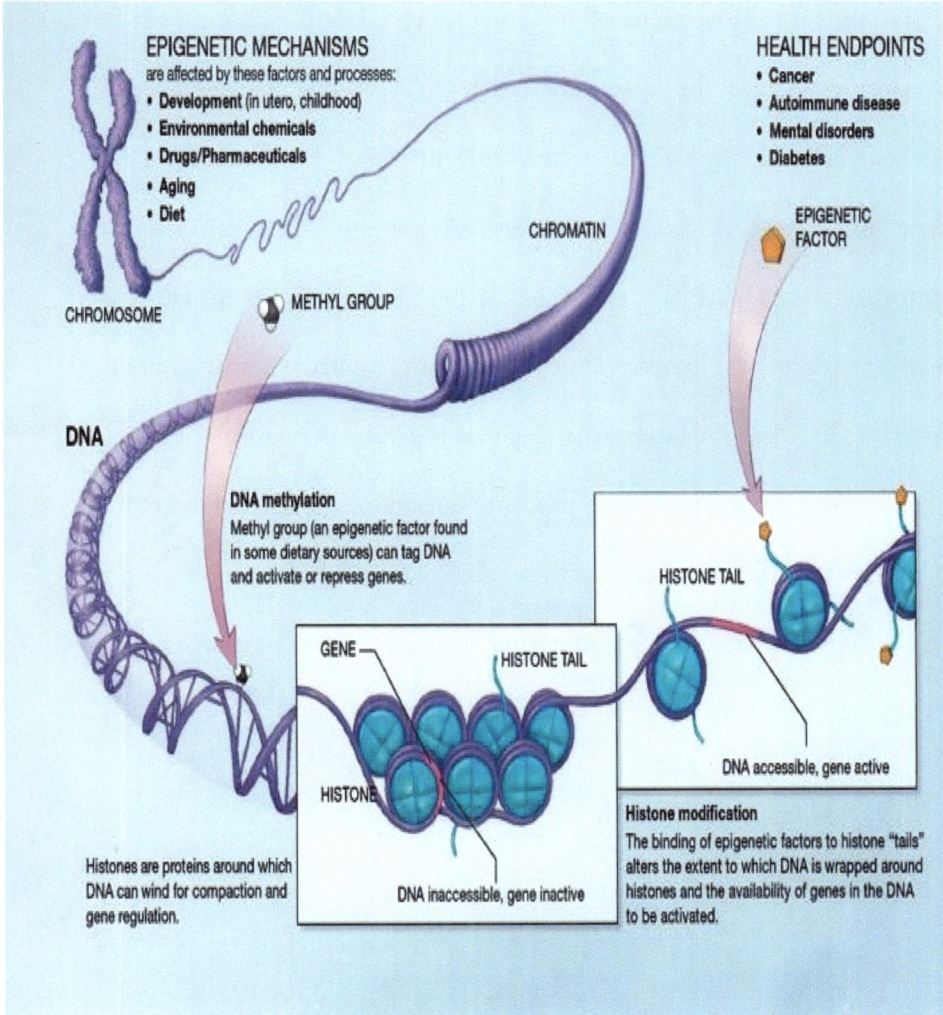

Figure 3 – Epigenetic Mechanism – The impact of environmental influences on DNA modifications at the molecular level. Epigenetics bridges the gap between the foods we decide to eat and the consequences of our decisions on our own DNA.

Although we now realize that DNA structure and function can be altered more directly by epigenetics, the DNA coding itself is for the most part beyond our immediate control outside of the laboratory. Of course there are exceptions within the labs, where DNA can now be manipulated to create a variety of clones and variants in animals. However there are major ethical dilemmas with altering human embryonic DNA in the laboratory and that is a subject beyond the scope of this book. Here we will focus predominantly on factors that everyone can control in their everyday lives to maximize their health. So let's begin with an overview of the environmental factors of health science in the next chapter.

Chapter 3

Environmental Factors

Environmental factors are critical components of health science, and play a major role in determining the outcome of the health/cancer equilibrium throughout our lives. The primary environmental factors include nutrition, exercise, stress reduction and the mind-body connection. We will delve into more details on these four critical environmental factors in later chapters. In the current chapter, we will simply begin with an overview.

The key point with environmental factors is that we have an abundance of control on them. It is our own decisions that determine our fate. We decide what to eat, when and how often to exercise, and whether or not we wish to be stressed out. With time, practice and patience we can discover for ourselves that we also have crucial controls over the mind-body connection. In fact, your body will not do much without your mind first telling it what to do.

As can be said of medicine, the same can be said of food – dosage is a critical factor. An abundance of consistent healthy choices will greatly improve your lifestyle and reduce your risk of cancer. If you want to be healthy in mind and body, you have to know about nutrition. Nutrition is a critical environmental factor that determines the science of our health throughout life. We will discuss nutrition in detail in a later chapter, however let's begin with some highlights here. For starters, remember to consume fruits and vegetables that are bright colors. Green vegetables are especially important. Fruits like blueberries can greatly reduce free radicals in the blood and minimize oxidative stress. Some other important healthy choices include regular consumption of foods like fish, nuts and whole grains. You also want to focus on low fat and protein rich meals to promote heart health, and reduce the risk of obtaining diseases like cancer. Poultry that is roasted, baked or broiled is best. Try to avoid fried chicken, as the hot oils in which it is prepared can lead to clogged arteries and potentially other diseases like cancer.

Regular daily exercise is a great way to meet your body's requirements to burn sufficient calories. Exercise promotes health and boosts the immune system to assist your body in fighting illness. Exercise promotes proper circulation of the blood in tissues thereby preventing an anaerobic microenvironment. Cancer cells thrive in anaerobic environments, so the oxygenation of the blood through regular exercise will greatly reduce your risk of getting cancer. Exercise may prevent DNA methylation of tumour suppressor genes, thereby inhibiting epigenetic mechanisms that could potentially be cancerous. Regular cardiovascular exercise can also clear blockages in your arteries thereby reducing the likelihood of heart disease.

Healthy eating and regular daily exercise can be an effective way to reduce stress in your life. Since stress reduction is such a critical component of living a healthy and fulfilling life, in a later chapter we will discuss additional options available to assist you in regaining control of this critical environmental factor in health science.

If we take it a step further, we come to realize, that many of the stress, problems and possibly even diseases in our lives originate or are sustained by the mind. The human mind is a fascinating and complex system. People have been studying the mind for centuries, and yet many of its functions remain elusive and abstract. However, numerous studies of patients recovering from illness simply through the power of belief suggests that modern medicine may be overlooking something critical here in patient care. As we will see in a later chapter, it is likely that the mind itself can be a vital determinant of the health of the individual. This is why it is critical to focus your mind on health, instead of disease.

So far we've highlighted four major environmental factors that are significant regulators of health science in the bodies of individuals, ultimately controlling their future state of health. In the next chapter, we will compare the similarities and differences between health science and cancer science. In subsequent chapters, we will delve into the four environmental factors in much greater detail. Then we will bring this information together with the health science guidelines. And finally, we will examine how combining these guidelines with the power of habit formation can literally change your life.

Chapter 4

Health Science vs Cancer Science

We introduced this topic briefly in chapter one (figure 1), however because of the common misconception that cancer science is a part of health science, let's discuss the real difference between these two sciences in more detail. Firstly, we may ask what does health science have to do with cancer science? In spite of a prevalent misnomer, common sense suggests that cancer doesn't have much to do with health science. We cannot have cancer and be healthy at the same time, so cancer and health must be separate events. At best, cancer exists at the far edge of health science, where health is beginning to fade (figure 4). Health science begins with genetics and environmental factors like nutrition and exercise. If something goes wrong in the human cells, health science can become cancer science. The goal for cancer doctors and researchers is to bring cancer patients back into health science. In a healthy population, the goal of each individual should be to never stray from normal health science in the first place.

Genetics	Genetics
Epigenetics	Epigenetics
Environmental Factors	Environmental Factors
Healthy Nutrition	Carcinogens
Regular Exercise	Malnutrition
Health	Lack of Exercise
	Ilness and Disease
	Cancer

Health Science   **Cancer Science**

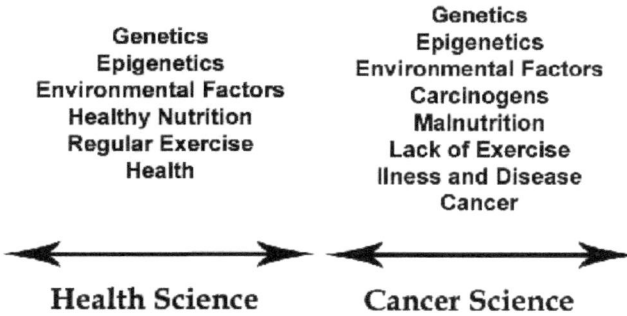

Figure 4 – Main differences and similarities between health science and cancer science.

Assuming genetics and environment are normal, the natural state of life is health. If we look at the animal kingdom, we see that cancer incidence is often far less than what has become commonplace in humans. Why is this? It's likely because animals are closer to nature than modern humans. Animals don't consume processed foods or smoke cigarettes, and they likely have an abundance of opportunity to exercise in the wilderness throughout their lives.

What we can take from this is that as we have progressed as a civilization we have grown apart from nature and moved towards industry. Obviously industry is a vital part of our modern lives, however if we cannot return to nature we must bring nature back towards us. For those that do, the rewards of health and prosperity await them in their elder years. However, those that neglect the nature throughout their lives must inevitably pay the balance towards the end. The price tag for consistent neglect is often diseases such as cancer.

So how then do we avoid this? How do we bring nature back into our lives and keep it? Nature includes all the environment that is natural and essentially creates the foundations of the health science guidelines. We will discuss these guidelines and how to implement them with habits later in this book. Let's now begin with the initial crucial components in achieving a healthy lifestyle: nutrition, exercise, stress reduction and the mind-body connection.

Dr. Kevin A. Figueiredo, PhD

Chapter 5

Nutrition Recommendations

Nutrition doesn't take much time, and it is critical to becoming a brand new healthy you. It will be much easier to consume nutritious foods when you know where to look. The ideas in this chapter can help you decide what needs to be added or removed from your current diet.

Good health requires proper nutrition, so it is important to make the correct nutritional choices. Select items that are made with wheat, oats, barley or brown rice. Oats in particular are a great way to lower your cholesterol naturally, keeping your heart healthy and preventing blockages in your arteries. Whenever possible, avoid white bread and flour products since whole wheat is so much healthier for you. The latter provides more fiber, and may potentially reduce the risk of several diseases including cancer, diabetes, stroke and heart disease. There are many nutrients that may be lost when your food is processed, so keep this in mind during your next trip to the grocery store. Don't be deceived by processed foods that look like good foods.

As you weigh your options for achieving optimal nutrition, it is important to note that not all foods that look healthy are as healthy as they appear. For example, a bowl of processed vegetable soup may not be as healthy as its unprocessed counterpart. As another example, ground turkey is a great alternative to ground beef, however it's fat content can actually be similar to that of ground beef depending on the cut. To maximize the health benefits of consuming turkey instead of beef, always look for the lean, extra-lean or low-fat choices while shopping for meats at your local grocery store.

Ensure that your food choices really are nutritious and healthy. Sometimes people may assume that they are selecting something nutritious at the grocery store when in fact, the opposite may be the case. Some foods are advertised as being healthy, but are often no better than the unhealthy counterpart. Oftentimes you can find these products side-by-side on the grocery shelves pretending to offer the customer a choice. To avoid this situation, minimize the time you spend in the aisles of grocery stores and always look closely at the ingredients of any processed foods.

The healthiest items are usually around the outer perimeter of the store anyway. Just as the body has daily needs for nutrition, there are also many prolonged health benefits in making the right choices every day.

It is also recommended to eat brightly colored foods. These are high in nutritional content, and because they are also low in calories, colorful fruits and vegetables have tremendous benefits. Focus mostly on the vegetables, and add a few fruits to improve the flavor. If you consume brightly colored foods with each of your meals, you will be on your way to a much healthier lifestyle. Many of the vegetable or fruit antioxidants reside in the skins, so it is important to consume the skins when possible. Try a wide variety of vegetables and fruits. Each contains a different combination of micro-nutrients that is essential to your body's daily requirements. An apple per day is a good start, however don't forget to add some vegetables like kale, spinach and carrots. You can also mix a variety of fruits and vegetables in a blender to produce delicious healthy smoothies. Add some water to a blender mix of vegetables like tomatoes, onions, mushrooms and kale for a potent cancer-preventing concoction that will keep you healthy throughout life.

The lycopene in tomatoes can prevent prostate, lung and stomach cancers. Onions can reduce the risk of colorectal cancer and ovarian cancer. Whereas mushrooms have been shown to reduce the risk of breast cancer. The antioxidants known as carotenoids and flavonoids found abundantly in kale have been shown to reduce the risk of a variety of cancers including bladder, breast, colon, ovary, and prostate. Regular consumption of kale can also lower your cholesterol levels, thereby minimizing your risk of heart disease. As an extra bonus, you could also add some slices of turmeric or ginger root which has been well established in promoting health and preventing a variety of diseases, including cancer. Since most cancers have many similarities at a molecular level, a vegetable that has been shown to have cancer preventing properties in one type of cancer will likely have beneficial effects of preventing other cancers as well. So by preparing smoothies of tomatoes, onions, mushrooms, kale and ginger root on a regular basis, you are giving yourself and your family a significant edge in skewing the health/cancer equilibrium towards health and away from cancer.

Of course not everyone is going to like the taste of all vegetables mixed in a blender. So it is fine to add some fruits for additional flavor. However, it is best to use at least two thirds of vegetables and no more than one third of fruits. This will ensure that you are maximizing the health benefits of these foods, while minimizing the risk of consuming too much sugar which is prevalent in fruits. Fruits are a great way of adding flavor, however they should be used in moderation. In contrast, vegetables may have less flavor and might not be as tasty for some people, however their nutritional content is fantastic. Vegetables are somewhat like bad-tasting medicine. The worst it tastes the healthier it probably is for you. Fortunately, if you really don't like the taste of some vegetables like kale or spinach, you can always mix them with fruits and still receive the health benefits.

Now moving on to what not to eat. If you always feel that you must consume junk food and sweets, ending that addiction can be one of the best things you can do for your body and health. If you regularly eat fast food, you might be addicted to the chemicals and synthetic substances often added to fast foods to improve taste and boost sales for the fast food chains. However, although this may be beneficial to the fast food restaurant's bottom line, it is not helpful to you in achieving the nutritional objectives required to live a healthy lifestyle. Because of the chemicals in these fast foods, you may be craving them even weeks after giving them up.

Make sure to always keep healthy alternatives around, so that when you get your next craving for french fries, reach for some celery sticks and carrots instead. Having healthy alternatives like nuts, fresh fruits and vegetables available constantly will make it easier for you to not think about unhealthy junk food and fast food.

In terms of nutrition, what you drink is just as important as the foods which you eat. Sodas are filled with high levels of sugar and chemicals that can have a significant detrimental impact on your health over the long term. Gradually replace your soda with water or tea. Adding lemon to your water will improve the alkalinity of your body, and assist with maintaining a healthier you. And don't forget your daily dose of a great vegetable and fruit smoothie.

When preparing recipes, many people like adding creams to their preparations. Even though it may make your food taste good and rich to you, it will also make your waistline expand. As a substitute, you could add silken tofu which has the creamy texture and some additional benefits such as additional protein. Something else to avoid is high fructose corn syrup. This substance is detrimental to your skin, and can also add fat and calories to your diet. High fructose corn syrup is also a toxin that can cause problems for your pancreas. It is often found in sweets which you should want to eliminate from your diet anyway.

As you age, you should try to reduce the amount of salt you consume each day. Often you may not realize how many salts you are consuming since they are found in so many different types of processed foods. It is suggested that you read carefully the labels of the foods at the grocery store before purchasing. High salt intake can increase blood pressure which can often lead to heart disease later in life. Preferably reduce your salt intake to less than one teaspoon per day, or as recommended by your physician. Avoid processed foods whenever possible, and don't add salt when cooking. A healthier alternative is to choose spices or no-salt seasoning mixes for flavoring your meals.

To keep your family eating healthy, bring your own meals or snacks for family outings. When indoors, it will also be easier for everyone to reach into the fridge to grab healthy apples and fruits, if they are readily available. This can be done by keeping them in a bowl, so they are easy for anyone to reach when they are ready for a snack.

Remember to eat breakfast to improve nutrition. Skipping breakfast will make you more likely to overeat later in the day. It can also increase your cravings for unhealthy foods high in sugar or fats. When your body is satiated with a good nutritious breakfast each morning, you are less likely to have cravings for unhealthy foods later on.

In today's society, nutrition is certainly on everyone's mind, as people become more concerned about how they feed their bodies and the impact on their health. For some, navigating the supermarket to choose the right foods can be a great undertaking, however it is much easier if you have the right information. As suggested earlier, while grocery shopping, try to spend most of your time on the outer perimeter of the store and less time in the aisles. The outer perimeter of most grocery stores usually contains the most nutritious foods, whereas the inner aisles are often loaded with processed foods.

Seafood is a great way to stay on a healthy meal plan. It has lots of quality nutrients including essential fatty acids and lean protein. So consider serving seafood for dinner on a regular basis. However, be cautious with the way it is prepared. Fish that is steamed or baked is best. Avoid seafoods that are deep fried in saturated fats. Reduce land based meats in your diet or replace them with fish, loaded with omega-3 fatty acids. Eating more fish will promote health for your body and your brain. Fish are high in Omega-3 fatty acids which improves vocabulary, memory and dexterity in nonverbal tasks. The risk of diseases like cancer and Alzheimer's can also be dramatically reduced by regular consumption of fish containing Omega-3 fatty acids. In addition, vegetarians that eat fish regularly have shown to have dramatically lowered risk of obtaining cancer in their lifetimes. Fish is a great source of healthy protein, and your heart will also benefit from the protective presence of the Omega-3 fatty acids.

Regardless if you are a vegetarian or meat eater, protein is a crucial component of your diet. To maintain healthy blood insulin and sugar levels, some protein should be consumed with each meal. If you eat too many carbohydrates and insufficient protein, you may experience a brief energy high, however your body will likely have to pay a premium for this, as an energy crash is likely to follow.

Vegetarianism has become very popular, and some vegetarians, known as vegans, choose to remove animal products completely. The foods consumed by vegans are often missing certain important nutrients like vitamin B12 found in many meats, or vitamin D found in milk. Vegans must ensure they are getting enough of these vitamins from natural plant sources or they should be taking the supplements.

They must also ensure they consume enough protein to maintain proper muscle mass. Quinoa is a great way to receive protein without consuming meat. It a superb source of essential amino acids, it is gluten-free and contains many great vitamins and minerals that are required by your body. Quinoa also has potent plant bioactive substances and anti-oxidants that can keep you healthy throughout life when it is consumed regularly.

For some tasty additions that are filling consider adding beans to your meals. They are critical in promoting the passage of foods through your intestinal tract, and they also have many important nutrients. Beans also have lots protein which makes it an especially great resource for vegetarians and vegans

If you have a craving for a salty or sugary snack, try eating unsalted nuts. Almonds, peanuts, and walnuts are low in calories and high in protein and vitamins. People who eat nuts are less likely to have heart disease and are more likely to live longer.

Vitamins and minerals are also critical components of a healthy diet, so ensure you are receiving them in sufficient quantities daily. Many food items in grocery store have been over processed and lack enough vitamins and minerals. If you are uncertain whether you are getting enough vitamins in your diet, consider taking a multivitamin to cover your bases. There are numerous important vitamins available in our foods and in pill format. We will highlight a few crucial ones here.

Vitamin C is important part of a healthy diet, as it is a very useful antioxidant. The collagen required for healthy gums and blood vessels is produced when vitamin C is available in the blood stream. This collagen is also required for wound healing and teeth and bone development. Vitamin C is potent as it helps with treating skin infections, acne, stomach ulcers, gum disease and colds. In addition, vitamin C has been found to decrease risk of cataracts, heart disease and cancer. Citrus fruits & some veggies are very high in vitamin C. If you don't consume enough, consider supplement.

A healthy diet also includes thiamin, also known as vitamin B1, which helps you utilize energy from carbohydrates effectively. Thiamin also assists with regulating your appetite. This vitamin can assist your heart, nervous system and the functions of your muscles. Thiamin is found in many foods, however it is not available in refined foods. This once again, highlights the importance of choosing a diet that favors unprocessed natural foods high in micro-nutritional value.

Riboflavin, also known as vitamin B2, is also an essential component in any healthy diet. It is required by your body to help release energy from protein, carbohydrates and fats in the diet, and it is involved in metabolism and transporting iron. Dairy products and whole wheat are rich sources of riboflavin.

Your body must synthesize numerous vitamins and minerals. Best to know which ones work well together when taking the supplements. Iron is an essential part of red blood cells. These cells transport oxygen in your body, so iron is especially important for competitive athletes. The body has difficulty absorbing iron with calcium present, so don't take iron pill with a glass of milk or antacid medication.

Vitamins are also an effective way of getting rid of muscle aches. Strained and overworked muscles can benefit greatly from fish oil supplements and vitamin E. Another great vitamin is vitamin A which is an antioxidant that enhances the immune system, improves vision and reduces heart disease. Squash, carrots and dark leafy greens provide natural source of vitamin A.

Manganese is a great mineral to consider. It speeds up healing process and contributes to metabolic process. Manganese can be found in nuts, grains, leafy greens, beans and tea. Zinc is also a useful mineral. Take a zinc supplement to help your body fight off the cold and flu. Zinc supplement can also prevent you from getting infections and other illnesses.

A simple and effective way to get started with healthy nutrition is to consider one of the best diets available: The pyramid diet presented in figure 5 is based on the Mediterranean diet. It offers you some guidelines on what constitutes a healthy diet. Start with plenty of vegetables, fruits and grains, and gradually work your way up to fish, seafood and poultry with each daily meal.

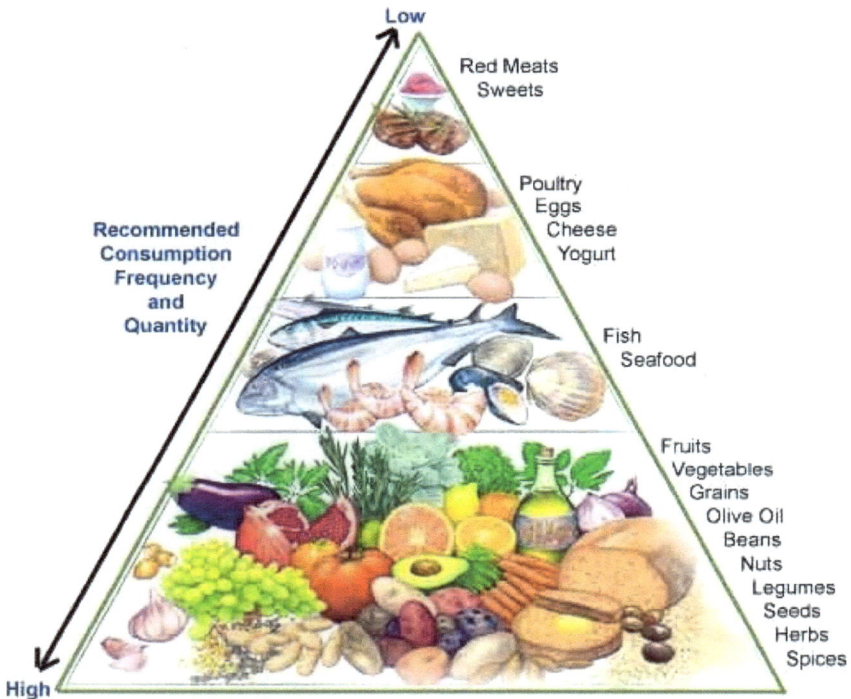

Figure 5 – The Pyramid Diet - One of the best diets available is based on the Mediterranean diet, where foods derived from plants are consumed in the greatest quantity, followed by fish and poultry products. Red meats and sweets are not required for a healthy diet, however they can be consumed in low quantities as long as the bottom three layers of the pyramid are adhered to consistently.

Healthy nutrition doesn't have to be overly complicated. Just keep it simple, and if you sometimes forget what is really good to eat, then just refer back to figure 5 to get you back on the right track to healthy nutrition.

To achieve optimal health, we must aim to become more careful and discerning with the choices that we select to fuel our bodies. There are plenty of temptations of unhealthy choices in the plethora of restaurants and fast food outlets in the cities. However, if you can consistently apply the information in this chapter to your everyday life while avoiding the distracting advertising of unhealthy products, you will be on your way to a healthier you. Plan your meals in advance on a regular basis, and you'll soon find that proper nutrition doesn't require much effort when it becomes routine. And the rewards of healthy living over a lifetime can be staggering, especially if you want to alter your future when the odds are most in your favor...

Before Cancer.

Dr. Kevin A. Figueiredo, PhD

Chapter 6

Exercise Recommendations

Health is very important to life, so exercise regularly to always keep fit. To maintain good health, one should have a healthy diet as we discussed in previous chapter. Perhaps even more importantly, one should do regular exercise to keep fit.

You can do many kinds of exercises at home. They can range from aerobic to anaerobic. The main thing to note in any exercise is that you should not overdo it. Just as overeating can be bad for your health, over exercising can also have detrimental effects. So be cautious with this aspect. There are different kinds of exercises for men and women. Similarly, you have different exercises for different age groups. The 'Physical Activity Guidelines for Americans' stipulates different durations of exercises for different age groups. We'll discuss how durations varies with age for men and women later in this chapter. However, first let us look at some of the benefits of regular exercise. We will also compare some types of exercises as well as the difference between anaerobic and cardiovascular exercises.

There are innumerable benefits of regular exercise. We will only highlight the most important here. Regular exercise can keep your weight under control and prevent you from becoming obese. Obesity is the cause of many an ailment today. Regular exercise can also dramatically lower your risk of diseases such as cancer. Physical activity can reduce the risk of many types of cancers including breast cancer and colon cancer.

A well-known benefit of exercise is that it can reduce the risk of cardiovascular diseases. Regular exercise can keep your heart muscles fit and lean. It will prevent buildup of fats. Hence, your heart will stay healthy and your blood vessels will remain unclogged. Exercise also reduces risk of ailments such as diabetes. Keeping an exercise routine can give a solid workout to every part of your body. Your internal organs also remain fit. A fit liver is a fantastic asset. A fit liver will not allow diabetes to be an issue for you.

Exercise can strengthen your bones and muscles. This will reduce the risk of fractures at a later stage in life when the bones tend to become brittle. Exercises will not only keep you physically fit but mentally alert too. It can improve your mental faculties and keep your mind fresh. You will be able to do your daily chores easily and comfortably. You can continue this habit even when in old age, and you will be able to retain your independence. In fact, exercise can help you live a longer life and a more satisfying life.

Let's now discuss the two main types of exercises, anaerobic and cardiovascular exercise. We will also see their benefits as well as differences.

Anaerobic exercise is an intense form of exercise usually done by sportsmen and athletes. It should be intense enough for formation of lactate. Usually athletes involved in non-endurance sports do anaerobic exercises. This can help them to build up strength, speed, and power. This can help such sportspersons to perform better in short duration races such as the 100 m dash and similar events. This exercise involves the strengthening of the fast twitch muscle, muscle fibers, as well as the heart muscles.

Exercises requiring maximum levels of exertion such as sprinting and weight lifting are some examples of anaerobic exercises. Even high intensity exercises such as cycling and rowing which are aerobic in nature become anaerobic when you perform these with more than 90% of your maximum heart rate. Additional benefits of anaerobic exercise include building lean muscle mass, burning calories effectively, reduced weight and it can build your endurance and fitness levels as well.

Cardiovascular exercise, also known as aerobic exercise is a physical exercise with a low intensity. The very name suggests that this exercise requires free oxygen. You would expect to be able to sustain these exercises for a long period. Types of aerobic exercise include jogging, swimming, and biking as examples. There are numerous benefits of aerobic exercise such as follows.

Firstly, aerobic exercise strengthens the muscles involved in respiration enabling a free flow of oxygen into your lungs. It strengthens the heart muscles and improves its pumping activity. It also improves the circulation of blood and reduces blood pressure. Aerobic exercise increases the red blood cell count and facilitates transportability of oxygen.

In addition, it improves the mental health of a person as well. Aerobic exercise also reduces risk of multiple diseases including cancer, heart disease and diabetes.

The main difference between aerobic and anaerobic exercises is the presence or absence of oxygen. Usually cells get their energy by using oxygen to fuel metabolism. As long as you exercise with adequate oxygen intake, muscles can contract without fatigue. This is the aerobic exercise. Once you increase the intensity of the exercises, muscles start relying on other reactions not requiring oxygen for muscle contraction. This activity can produce waste molecules that can impair muscle contractions. We call this muscle fatigue. Aerobic exercise is a low intensity exercise done for building endurance and stamina. If done correctly in moderation, you will not suffer from muscle fatigue. This does not result in production of lactic acid. By contrast, anaerobic exercise is a high intensity exercise done for building speed, strength, and power. In this case, you can suffer from muscle fatigue due to reduced supply of oxygen, as you will tend to produce lactic acid in significant quantities.

Different exercises require different intensities. At the same time, the intensity of the exercises depends upon the age and gender of the person too. The Physical Activity Guidelines for Americans recommends that a normal adult should exercise for at least two and a half hours per week. Some people may need more and some may need less. However, the optimum duration should be about two and a half hours per week for one to maintain his or her weight within permissible limits. You should always start your exercises at a lower intensity rate and build up gradually. Children as well as adolescents should have at least one hour a day of physical activity. This could range from simple activities such as running and playing to a more organized form of aerobic exercise as well. This can help them to strengthen their muscles as well as bones.

Healthy adults should spend about two and a half hours on exercise on a weekly basis. In case they tend to prefer anaerobic exercises, they can do so for one and quarter hour per week. You can start with a brisk walk for around thirty minutes on a daily basis for about five days a week. You can supplement the walk with an alternate day of jogging as well. Women can also do jogging as well as cycling.

Other types of exercises common among women are skipping, doing sit-ups and squats. Women should not generally do weights without consulting a professional.

Let us examine the duration of regular exercise based on age (figure 6). For simplicity, we will divide people into four age groups. (1) Children under 5 years of age, (2) Young children and adolescents between 5 years and 18 years, (3) Adults between 18 years to 64 years, and (4) Older adults over 64 years.

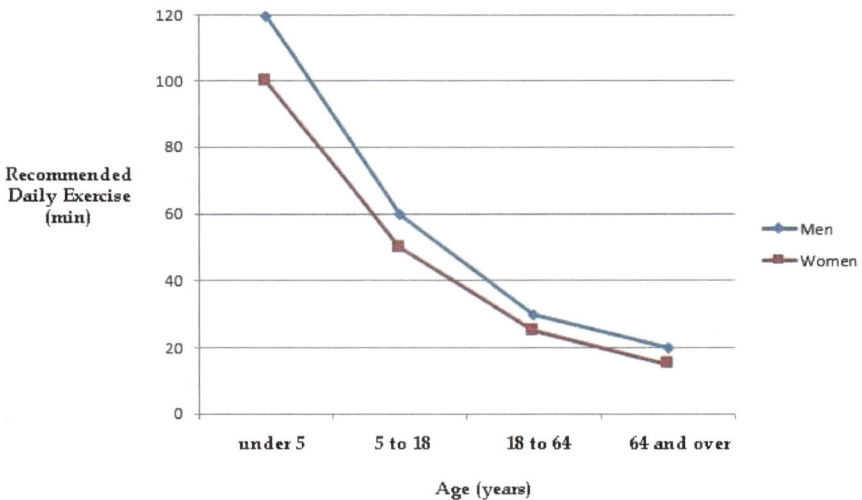

Figure 6 – Recommended daily exercise varies with age and gender. In general, as we age the amount of exercise will decline for most of us. However if we transition towards increasing cardiovascular exercise instead of anaerobic exercise, we can continue to receive the tremendous health benefits of exercise right into our twilight years.

Throughout life, our day to day physical activities will vary substantially. Here we'll divide activities into two groups: moderate and strenuous. For maximum health benefits, alternate between both throughout life. Much will depend on your existing level of fitness, and how well you have taken care of yourself up to the current stage in your life. Listen to your body closely as it will tell you which type of exercise is best for you.

Let's start with group 1 from figure 6. Babies should be active throughout the day. You should not allow them to watch television for long periods. Babies should do regular exercise for at least three hours a day, and you can spread this out between indoors as well as outdoors. Some activities that can keep babies busy include standing up, moving around, walking and simple games. For toddlers and other children under 5, slightly more strenuous activities can be done, including active play, running, riding a bike, skipping and gymnastics.

In group 2 from figure 6, there are young adults. Young adults should do simple exercises for at least an hour daily. This can be moderate to intense activities. Moderate activities include walking to school, skateboarding, rollerblading or walking the dog.

More strenuous activities for your adolescent may include dancing, running, cycling, swimming and martial arts.

The group 3 in figure 6 covers the vast majority of adults from age 18 to 64. This is a large group, and there will be much variation among individuals, however in general we can expect the intensity and duration of exercise to become less as we age. As an estimate, adults should have a routine of minimum two and half hours per week or at least half an hour per day. Some moderate activities for adults include walking, aerobic exercises, dancing, canoeing and volleyball. More strenuous exercises include running, swimming, playing tennis, football or hockey.

In group 4 from figure 6, there are older adults. These individuals should reduce the workload by reducing the exercise period to about a hundred and thirty minutes a week. This might work out to around twenty minutes a day. Moderate exercises for this group include walking, gardening and lawn care, or bike riding on level ground. Exercises that may be considered strenuous for this group would be hiking uphill or aerobics.

In general, individuals that begin regular exercise routines early in life and maintain it, often see a divergence in mid-life between themselves and their peers who do not exercise regularly. For those that exercise regularly throughout life, chronological age becomes much less important than biological age. This divergence explains why individuals in their 70s that have exercised throughout life may sometimes look no older than 50 or 60. The important thing to remember about exercise is that it has to become a habit, and it must be done regularly in order to experience the greatest benefits later in life. So we have seen the importance of exercise throughout life, and we have also seen the difference between aerobic and anaerobic exercises. Everyone should have a regular schedule of exercise, no matter how old he or she is. Regular daily exercise will enable you to live a healthy as well as a cheerful life.

Chapter 7

Stress Reduction

Stress is damaging to your mental and physical health and is a major factor in many diseases including heart disease, diabetes and cancer. So let's discuss this important foe to healthy living in some detail in this chapter. Healthy nutrition and exercise, as well as understanding your mind-body connection will be a great starting point in eliminating stress in your life. This chapter can help you feel much better about what causes your stress as you discover additional ways to deal with it.

Stress triggers the release of adrenaline and cortisol which promote the fight-or-flight response that evolved in humans thousands of years ago. Stress was very useful in human evolutionary past since the triggering of this fight-or-flight response allowed our ancestors to escape extreme dangers such as an attack from a tiger or lion.

In the modern world, these extreme dangers are seldom present so stress has become a somewhat unnecessary relic of the past. The fight-or-flight response is still useful today in some situations, as it may allow you to survive a car accident, and it may assist if a bear charges at you when hiking. However, for most of our lives, stress is not very useful. Nonetheless, it persists in our lives, but only if we choose to allow it.

Before we can conquer stress we first have to know what causes it in our lives. A great way to determine what causes your stress is to begin tracking the way you react during stressful situations in your life over several weeks. Monitor how you respond to these stressful situations, and then look for ways to change your responses going forward. Looking back at your responses will enable you to determine if your stress was healthy and productive. And if it wasn't healthy, then consider changing your response to it. Often we cannot control the cause, however we do have complete control of our responses so be sure to use that to your advantage.

When you realize the impact that stress is having on your life, the next step is to find ways to minimize or eliminate it. There are plenty of great ways to do this, so let's discuss some of the most common.

Releasing anxieties and also other emotions is an essential method to help yourself feel a lot better. Yoga and meditation can also be a tremendous benefit in reducing the stress in your life. In addition, having animals around can decrease your stress level. Research has shown that the act of stroking an animal's fur reduces levels of stress within a matter of minutes. Conversing with someone near to you can be a good idea for relieving stress. If you think you are stressed, try rubbing a few drops of spearmint oil on your neck and temples. Forgiving others can also ease your life. Another great way to relieve stress would be to drink a good cup of hot tea. There are lots of different types of tea that assist with stress, including herbal, chamomile and kava kava. Steep the tea for approximately 10 minutes to achieve the most from the herbs. At night, in order to reduce stressful feelings of the day, drink a cup of tea to help you sleep.

Music is also definitely an effective component of an excellent stress relief plan. It is actually a recognized phenomenon that music therapy can lessen stress provided that the kind of music is soothing to you. Music therapy facilitates relaxation and induces the brain to produce serotonin. Although be careful with the music selected because listening to some modern rock stars and popstars lyrics may actually have the reverse effect. For this reason when you are trying to destress, instrumental music without lyrics may be your best choice.

Managing your time and efforts can also result in decreased stress. As you manage your time better, you will find that your stress becomes more manageable. Staying on a schedule and thinking ahead with effective time management, can allow you to get far more done while avoiding stress and overwhelm. Learn how to say no when you are overwhelmed by stress and becoming stressed out. Trying to do excessive work will only result in increased levels of stress hormones in your body, no matter how much you might like to please others.

In addition, self-hypnosis has proven itself to be an efficient treatment for individuals who are disturbed easily from unimportant events. Your body often tightens up whenever you feel stressed. This includes the shoulders, back muscles and jaw muscles.

Once you have identified areas that are tightened by stress, make an extra effort to stretch those parts of the body to dissipate that stress. Stretching these muscles will relieve tension and allow you to more easily relax. Shallow breathing can also add to the feeling of stress because heartbeat may increase and your chest muscle may begin to tighten. So take some deep breaths to bring more oxygen into your bloodstream which can help you relax quite quickly.

Spend some time in your life doing things that you enjoy. It will help you endure, when the anxiety comes your way. When you find a good hobby or activity that you enjoy, make certain you devote time and energy to it on a daily basis. Doing something you love at least once every day will dramatically increase your health and sense of well-being over time. So tune in to music, take a relaxing walk or sit back and read an excellent book. In order to prevent stress overload, even if you are very busy, it is important to enjoy your favorite activities. If a certain circumstance is causing stress in your life, take yourself away from the particular circumstance. If you have an important presentation or meeting coming up, instead of getting stressed about it, tell yourself that you are enthusiastic and looking forward to it.

This will make it easier to stop worrying about how you might have been feeling.

Another great method to reduce stress is to visualize something which calms you. Take a few moments to really take into account the things in your daily life which may have made you happy and restful. Don't put things on the list without making an effort to visualize them in great detail using all of your senses. It is very important do this daily, and it can help immensely whenever stress-inducing thoughts occur. Don't permit the troubles around the globe to weigh too heavily upon you. If negative news broadcasts are affecting your daily mood, all you have to do is stop watching TV. Attempt to behave like you feel positive and that all is well within your world.

If you are very stressed out, you may be tempted to indulge in a dessert or a treat. This is not recommended when you are stressed out. Desserts are fine once in a while, however eating these when you are stressed out shows a lack of self-control. Getting rid of the stress in your life should be your first priority. Best to indulge in desserts and sweets when you are stress-free. This will make it much easier to keep your healthy lifestyle on the right track.

When you are stressed out and eating desserts, you greatly increase your risk of overeating. Once you have properly assessed the level of stress in your life, remember the chapter on nutrition recommendations before considering this option.

Stress may also be an outcome of making bad decisions and an inability to accept any kind of responsibility for them. Taking responsibility for your own decisions is a great way to reduce stress in your life. Some may wish to even take this a step further by assuming responsibility for everything that happens in your life. This method can dramatically reduce your stress levels. When you take responsibility for everything that happens to you, you maintain control of your own life and cease to become a victim of circumstances beyond your control.

The consequences of ongoing chronic stress can be extremely detrimental to your overall health and well-being. Chronic stress can actually change your physiological responses and make you vulnerable to several diseases including cancer. Hopefully the ideas in this chapter, can assist you with reducing or eliminating stress from your life. Work on stress reduction daily, and eventually you will find that you can defeat stress in your life once and for all.

Dr. Kevin A. Figueiredo, PhD

Chapter 8

The Mind-Body Connection

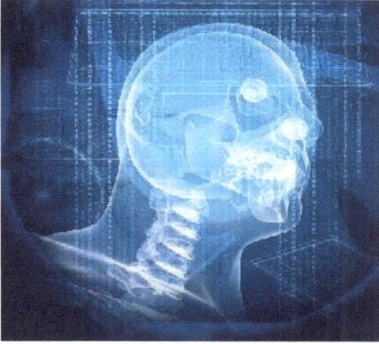

The mind-body connection is vital in living a fulfilling and healthy life. Although this topic is currently not fully understood by modern science, it is a fascinating subject that can have a major impact on our lives. The mind-body connection cannot be fully quantified and measured at this time, and experimental evidence of this subject is limited. Science requires repeatability, reproducibility and rigor before any hypothesis can be considered factual, so lack of this evidence at the current time excludes mind-body connection from modern health science.

Nonetheless, it is included in this chapter because it is highly likely to be a critical component of future health science once human-made technology is better able to quantify the substantially more complex human-intrinsic technology that exists in each of us.

The importance of the brain's role in determining physiological outcomes is highlighted by the placebo effect. Placebo is an inactive substance which has no intrinsic value of its own, and it may create a positive effect when the patient believes it to be a real drug. The entire effect of the working placebo revolves around the psyche of human mind. Just by letting the mind think that it is actually receiving the drug, the mind convinces the body which may begin to heal itself. Placebos can be sugar pills or saline injections given to patients without their knowledge. Recently it has been observed that some placebos work better than other placebos, and this once again reflects the power of the psyche in regulation of the body's recovery. Colorful pills are better than white pills. Large pills work better than smaller pills. Injections have more promising effects than oral pills. And surgeries work more efficiently than injections. So essentially most of the response is determined by psyche.

As an example, recent studies of knee transplant divided into groups showed that recipients that simply received surgical incisions reported the same recovery as those that actually had the knee surgery. This implies that the mind can heal the body on its own, at least in some circumstances. Placebos can also be used with classical conditioning technique in which a placebo is used with a real stimulus. In this manner conditioning and expectation together give a better outcome. In spite of these advances, the placebo remains controversial in medicine since it often introduces deception between doctor and patient.

Nonetheless, there are countless examples of patients believing they will recover from their illness, and often they do. The mind is a powerful force as it can naturally release your intrinsic pharmaceuticals at exactly the right dosage and at exactly the right time to help you fight disease. However, to properly regulate production the natural hormones and cytokines that can help you fight disease, you have to ensure that your thoughts are really your own.

As an example, if person A tells person B that he looks ill, and if person B does not have strong convictions and belief in his own intrinsic health, then this mere suggestion from person A may cause person B to become physically ill. On the other hand if person B, has a strong belief that he is healthy and strong, then the same suggestion from person A will not have much effect on person B. Your mind can help you heal and often it's just a matter of believing in yourself and having a positive outlook on your life. If you have a positive approach to life, every hardship can be a blessing in disguise.

It has also been observed the placebo effect, has a detrimental counterpart: the nocebo effect. The latter occurs when the administration of a placebo results in a negative effect on the patient. This may occur if the patient is inclined to believe that something will go wrong with their treatment or surgery. These types patients are often strong believer's in 'Murphy's Law'. For patients experiencing a nocebo effect, no matter the medication or saline pills they receive, oftentimes their psychological negative response can prevent recovery and may actually make their symptoms worse. This underscores the importance of maintaining a positive mental state even if you are ill.

Although additional studies are required to fully comprehend its usage, there is no doubt that the placebo effect when used in a positive manner can provide significant beneficial effects on the patient especially when administered by the patient themselves. It is important to note that the placebo effect may vary greatly within individuals, and may be entirely dependent on willpower and conviction of the individual. This is why if it is used at all, it should only be used as a supplement to conventional therapies with the approval of your physician. Until further studies are performed that clarify its impact, the placebo effect should not be used as a substitute for conventional treatments of any disease.

It is clear that the placebo effect has a dramatic impact on our lives every day. However, since this effect is controlled entirely by your mind, when you fully comprehend its usage, you will not require a saline pill to benefit from it. To improve our lifestyles we have to take advantage of this effect by using it to the maximum especially when you are healthy. And if you are ill, you have to believe that you are getting better. Ironically, the placebo effect is likely related to the concept that what we believe to be true often becomes our reality.

The brain is a complex biological system, and it consists of almost all pharmaceuticals that the average person needs throughout their lives. Furthermore, it produces these pharmaceuticals free of charge, at exactly the right time, and at exactly the right dosage. For individuals that are exposed to many types of medications for minor ailments, or for those that are convinced that a certain medication will benefit them, the placebo effect may possibly be most relevant here. In the end, the placebo effect may simply be the trigger that restores normal function of the natural brain pharmacy.

If we are able to control our thinking, change the way we perceive things and add positivity to our thoughts, we can promote our own health and healing. A positive outlook on your life will help you deal with difficulties better than someone whose expectations are less positive. A happy and a satisfied brain produce substances like dopamine which are natural painkillers. The brain also secretes endorphins which help relieve pain and strengthens the body's immune system. In essence it makes our exterior stronger so that it is less susceptible to negative influences. If used wisely, your mind can be your body's armor.

As we reflect on the impact of the placebo on the mind, we realize that there is something we have not yet grasped in this equation. Many people think of the mind as the brain itself, however there may be something else to it that more closely approaches the realm of the spiritual. In addition to the conscious mind which controls the vast majority of our thoughts, there is also a subconscious mind that may be even more critical in determining your state of health and well-being.

In contrast to the conscious mind, the subconscious does not analyze or evaluate anything. The subconscious mind accepts what it is given without questions. Think of the conscious mind as the guardian of your subconscious. Now this can work in your favor or it can work against you, dependent on your life experiences. Nonetheless, you may be able to bypass your conscious mind and communicate directly with your subconscious by consistent and repetitive suggestions of health and well-being. This can be done through the power of habit formation, and we will explore this topic further in a later chapter.

In the next chapter, we will review the critical environmental factors that determine the fate of your health, and we will summarize them as the health science guidelines.

Dr. Kevin A. Figueiredo, PhD

Chapter 9

Health Science Guidelines

These suggested guidelines may assist you and your family in avoiding the future impact and consequences of getting cancer and other diseases. Please note that they are only guidelines. Your results will vary and be dependent mostly on how strictly you adhere to these guidelines throughout life. The hope is that these guidelines will assist many who adhere to them, however even among those who follow these closely, it is important to remember that cancer is a very complex disease and there are factors that may be beyond your control if you do get cancer. All each of us can do, is try our best to minimize the factors that are within our control. And so with that, let's begin the guidelines....

1.   **Maximize consumption of micronutrients** by eating mostly bright and colorful fruits and vegetables daily.   Micronutrients are essential nutrients that are indispensable to your body's health. Micronutrients are trace minerals and vitamins that are often found in abundance in fruits and vegetables.   This last sentence looks like an oxymoron at first glance, however if you read it again, its significance will become apparent.   Give preference to lots of dark leafy greens, like kale and spinach. Also regularly add some nuts and seeds to assist with the digestion and absorption of the micronutrients from vegetables and fruits. Following this guideline will have a tremendous impact in nourishing your body and immune system throughout life.

2.   **Consume mostly low fat and protein-rich meals.** Fish and poultry that are baked or broiled are a great way to meet this requirement. Your level of commitment to this rule will have the greatest impact on the extent to which this will assist in maintaining unclogged arteries and veins in your body throughout life.

3. **Adhere to the food pyramid based on the Mediterranean diet** (figure 5). This is likely the best diet available to humans at the present time, and the degree to which you adhere to this diet or similar diets will likely be a significant determinant of your health later in life. In the end, the best medication is not a pill you take after you are ill... the best medication is the food that you eat everyday, and depending on dosage and quality it can either lead you to sickness or health. If you prefer health, then follow the food pyramid based on the Mediterranean diet.

4. **Exercise.** There is no substitute for exercise, so make sure to make it a regular part of your daily routine. Sign up at your local gym or fitness club, or ensure that you spend plenty of time outdoors daily. Regular exercise throughout life will have tremendous benefits on your health especially when you reach your elder years.

5.   Avoid unhealthy toxins, drugs and junk food. Even if cigarettes, tobacco and alcohol are legal drugs, it is important to remember that their chemical composition and biomedical definitions as drugs are not altered by government legislation. Tobacco smoke is especially dangerous as it consists of hundreds of potential carcinogens, any one of which can lead your body towards a cancerous condition. Junk food also has many chemicals that are addictive and can be hazardous to your body when consumed on a regular basis. So avoid junk food and drugs, including tobacco smoke and alcohol.

6.   Minimize stress in your life. Stress is a major cause that can lead to many diseases. Chronic stress often causes chronic inflammation in the body, which in turn may shift your health/cancer equilibrium towards an unhealthy cancerous state. So it is vital to minimize or eliminate detrimental stressful factors in your life.

7.   **Center your thoughts on health**, instead of disease. Consider the mind-body connection, and the impact of the placebo effect. Disease cannot exist in the presence of health, so it is always best to focus on health instead of disease. It is much easier to convince your body that you are healthy through the placebo effect, if you have followed the other health science guidelines throughout your life. The earlier in life that you get started with following health science guidelines, the greater benefit you will receive in your elder years.

8.   **Establish good habits.**   This is a critical final step that permits all the previous guidelines to function cohesively over time. Without this last step, the guidelines will likely not be useful. Habits are required to consistently follow these guidelines day after day, and week after week throughout your lifetime. You must instill habits into your lifestyle so that following these guidelines becomes routine. The better able you are to follow this final crucial step, the more likely you are to skew your health/cancer equilibrium towards health and away from cancer throughout your lifetime. Because of the importance of this final step, we will discuss details in the next chapter on how to form life changing permanent habits.

Dr. Kevin A. Figueiredo, PhD

Chapter 10

The Health Science Habit

The wisdom of the Greek philosopher and scientist, Aristotle, provides one of the key reminders to the entire human race that success does not come overnight: "We are what we repeatedly do. Excellence then, is not an act, but a habit." According to Aristotle, it is consistent discipline that allows us to move or progress from one point to another. This prompts queries for the ages: how exactly are habits, or else, discipline initiated? What factors are vital or contribute to building these habits? And more importantly, how are permanent life changing good-habits formed?

Given the countless diversions of the modern world, habits can sometimes be difficult to create. Daily livelihood comes with multiple distractions that can lead some off the path of good habit building, and back to the same old lackadaisical ways. To eliminate some of these difficulties in good and permanent habit building, and to assist you with fully implementing the health science guidelines from the previous chapter, let us begin a discussion on the current research available regarding motivation, habit building and discipline. We will then discuss some actionable steps through which permanent good habits can be established in your life.

Before proceeding into outlining the steps for permanent habit building practices, it is important to clarify the differences between short-term or motivational habit building practices and the long-term and the permanent habit building practices. First off, short-term habit building practices are based on motivation. Whereas the long-term habit building practices are based on discipline. Therefore, the most appropriate ways in distinguishing these two habit building procedures is to identify the practical differences existing between motivation and discipline.

On one hand, motivation is situational. This implies that motivation is based on the current situation of a person including the environment, mood, and personal precept. Also, motivation is fleeting is such a way that it comes and goes with no given consistent pattern. It might not last a month, a week, a day, or even an hour. It may drive one into beginning an act or even continuing it for some time. Inevitably, motivation could disappear at any moment, and is likely not sufficient in reaching the completion of one's objectives in the long term.

Motivation is everywhere. A person will pursue acts of attending a conference, reading newspapers, watching news and programs, and even attend counseling sessions only in search of motivation. However, there is no binding factor between motivation and action. Motivation to eat healthy, to quit drinking or smoking, and to exercise is often temporary especially if we've been doing these things throughout our lives. So what we have to do instead is form habits. The earlier in life we get started with habits, the more difficult they will be to break later in life. Here again we have to be careful because habits can either be good or bad.

If we maintain good habits of eating wholesome healthy foods, exercising regularly throughout life and following the health science guidelines, then we can dramatically reduce the risk of ever getting diseases like cancer, diabetes or heart disease.

In contrast to fleeting motivation, life changing permanent habits are created by self-discipline. This consists of features that primarily bind a person into doing a given action. For instance, discipline is a sample of permanent habit. It is consistent and habitual. Otherwise, with no consistency, it can no longer be referred to as habit. Therefore, it must occur repeatedly, just as Aristotle affirmed. How then can discipline be formed? How is this aforementioned motivation translated into a lasting habit?

Clearly, motivation works effectively in the short-term. With an existing goal, it is possible for a person to summon up enough motivation to pursue a task intently for a week or so. In cases where the goal is very important, it is possible to carry on the motivation uninterruptedly for up to a month. However, the motivation eventually wanes. If the goal at hand carries on for more than a single month, then the pursuer will need more than just motivation.

Discipline and habits then come into play. Luckily, the habit binding techniques are great since they translate the existing short-term motivations into a more durable discipline. Therefore, permanent good-habit building techniques will comprise of all those procedures that oversee the routine translation of motivation into discipline. Habit binding techniques may require one to invest in consistent routines using triggers, punishments and rewards in order to stabilize the motivation into a systematic habitual output.

There are several steps to permanent good-habit building as determined by recent studies in psychology. The first step in habit building is by creating micro-quotas and macro-goals. In other words, take small daily action steps, and dream big. Keep your dreams consistently in front of you, however take action on achieving your daily goal. By constantly reminding yourself of your dream you are creating an intrinsic motivator. The latter is what results in creation of habits that stick. This technique may be an effective method to building personal discipline and may assist you greatly with following the health science guidelines.

It is somewhat like driving across the North American continent. You might be completely and totally lost, but you know that if you keep driving in one direction, eventually you are likely to reach an ocean.

The second step involves creating behavior chains. This step ensures the creation of permanent habits by manipulation of existing routines, rather than trying to act against them. The procedure is built around the concept of if-then planning. It allows a person to check on some existing environmental or situational triggers to know that it is time to act out on a habit. The procedure can also be referred to as implementation of intentions, since it involves picking on an existing part of one's schedule and then adding a link to a new step or action. The final result includes a new link to the normal chain or routine or simply a new habit.

The third step includes eliminating excessive options. This may be the best plan towards maintaining long-term discipline. The procedure requires one to clearly outline the aspects in life considered as mundane and to make them as routine as possible. By doing so, it allows you to make fewer unimportant decisions. The steps to routinize must ultimately change one's schedule and environment.

The final step in creating permanent habits includes process plan. Unfortunately, most people skip this step and instead fantasize on what their life will be like after building the habit. Process planning, although a seemingly small detail, is certainly a very crucial step in the whole habit building structure. Planning gives the habit pursuer a chance to re-think again why they need the change implemented. As a matter of fact, recent studies confirm that mistakes affecting a permanent habit are often caused by what the pursuer visualizes before the habit actually forms. Accordingly, persons who include the process planning by visualizing the steps required for achieving a goal are more likely to remain consistent in obtaining the desired habit as compared to their peers who do not visualize. The visualizations work for two reasons which include planning and emotions. Planning visualization helps anticipate and focus attention on the individual steps required in reaching the goal, whereas with emotions, visualization allows the pursuer to clearly outline the individual steps resulting in reduced anxiety. Therefore, process planning simply binds the other three steps, and hence results in affirmed buildup of permanent discipline and habits.

To summarize this chapter, it must be emphasized that motivation is only temporary. Motivation is great for athletes who have to get pumped up for a big championship game. But guess what happens after the game?... So motivation is great for temporary adrenaline rushes, but for long term success day after day, and week after week, something else is required. That's where habits become important. Good habits are intricately intertwined with self-discipline which is the ability to make yourself do what you should do, whether you feel like it or not.

Chapter 11

Conclusion

Now that we know that cancer is a genetic disease triggered by environmental effects, we can take the necessary precautions to avoid getting cancer in the first place. As mentioned in my previous book, *Terminate Cancer*, the best cure for cancer is most likely prevention. The immune system plays a critical role in determining the balance between tumour growth or destruction in the context of cancer science. In addition, the immune system also plays a critical role in determining the health of normal cells throughout life in the context of health science. Remember it is the loss of health within your cells that can lead to cancer.

The immune system will be your greatest ally in protecting your health. However, if you fail to supply it with proper nutrition, or neglect to maintain a healthy mind and body through exercise, stress reduction and understanding of the mind-body connection, then it is possible that the immune system can turn against you and start taking signals from a chronically inflamed microenvironment which may eventually become a tumour-controlled microenvironment.

So it is vitally important to keep your immune system happy throughout life by nourishing it with healthy fruits and vegetables and some fish and poultry. Exercise regularly and maintain a healthy mental state to maximize the benefits of the mind-body connection. Your immune system responds to your thoughts as well as to the foods you consume, so treat it well and it will protect you from disease. It is likely that the immune system evolved in humans through a symbiotic relationship where it protects you, only if you treat it well. Similarly the genes in most cells of your body will be using the epigenetic interface to respond to your decisions on health. So treat all your cells well, and they will help you live a life of health and well-being.

The objective of this book, *Before Cancer*, as a prequel, has been to re-emphasize this very important idea: prevention may be the best cure for cancer. In actuality, scientists do not like to use the word 'cure' especially for a disease like cancer, simply because there is not just one cancer – there are potentially thousands of types of cancer. Each cancer is often associated with the unique genetic code of individual patients. This is why genetic profiling will become more critical in diagnosis and treatment of cancer patients in the future.

For many, cancer may become a disease with which elderly pass away, instead of a disease that causes them to pass away. So for those with the disease, long-term monitored treatment may be the prescription to improve survival rates. Note that this is not a cure, however it would be a great alternative for those with the disease. For those without the disease, *Before Cancer*, a cure is indeed possible and it is called prevention. Prevention encompasses many of the lifestyle choices that we have discussed in this book. Let's do a quick review of the details.

Health science begins with genetics and nutrition and continues with exercise, stress reduction and the mind-body connection. We have limited control on our genetics, however we have an abundance of control on the environmental factors. By combining these factors and using them effectively you will benefit not just from an additive effect, but rather a potentiation effect on your health. For example, eating healthy food and exercise could make you multiples of times healthier than choosing just one of those two items. Multiply in the effects of the other environmental factors, and you will begin to see dramatic health benefits in your life.

Once you have taken care of the main environmental factors, begin to review the health science guidelines regularly. The chapter on the health science habit will assist you with making these guidelines a consistent and binding part of your life, but only if you choose to use them. If you decide to use them, then make them a regular part of your life to maximize the health benefits that you will receive. If you are healthy, focusing on health will help you keep it. Similarly, if you are ill, focusing on health may help you regain it.

For most cancer patients, *Before Cancer* there is health. And cancer science may not even be necessary for most, if we only would follow the proper guidelines of health science. By applying the power of habit to the health science guidelines discussed in this book, you will be ensuring that you and your family live long and healthy lives. Use this book *Before Cancer* as your guide to remain within the health boundaries of health science.

BEFORE CANCER

Dr. Kevin A. Figueiredo, PhD

About the Author

Dr. Kevin Figueiredo is a PhD scientist with over fifteen years of experiences in cancer research and health science. He obtained his PhD from the University of British Columbia, Canada followed by Postdoctoral studies at Harvard University, USA.

In 2014, he published *Terminate Cancer* to describe a model of viral infection and immune response as a potential means to treat cancer. It outlines some alternative therapies that are becoming available to cancer patients, and it provides the reader with a background analysis of the emerging field of cancer immunology.

In 2016, Dr. Kevin Figueiredo's latest publication *Before Cancer* describes how health science and a better understanding of the science of health can help all of us minimize our risks of getting cancer.

Dr. Kevin Figueiredo established Doctor-Kevin.com to provide a platform for expression of insights on cancer research and health science. Doctor-Kevin.com motto: "Disease cannot exist in the presence of health, so let's focus on health instead of disease."

Contact Information:

Dr. Kevin A. Figueiredo
720 King St. W, Suite 416,
Toronto, ON
Canada M5V 3S5

email: info@doctor-kevin.com
website: **www.doctor-kevin.com**

BEFORE CANCER

Terminate Cancer

by Dr. Kevin A Figueiredo, PhD

Terminate ~~Terminal~~ **Cancer**

A model of viral infection and immune response as a potential means to treat cancer

Dr. Kevin A. Figueiredo, PhD

ISBN-10: 1514843234
ISBN-13: 978-1514843239

A molecular scientist specializing in immunology, molecular biology, genetics, and cancer research advances a theory of cancer treatment based on induced immune response.

"Evaluating the efficacy of this proposed treatment method is the purview of cancer doctors and researchers... Figueiredo may be on to something."
—*Kirkus Reviews*

www.doctor-kevin.com

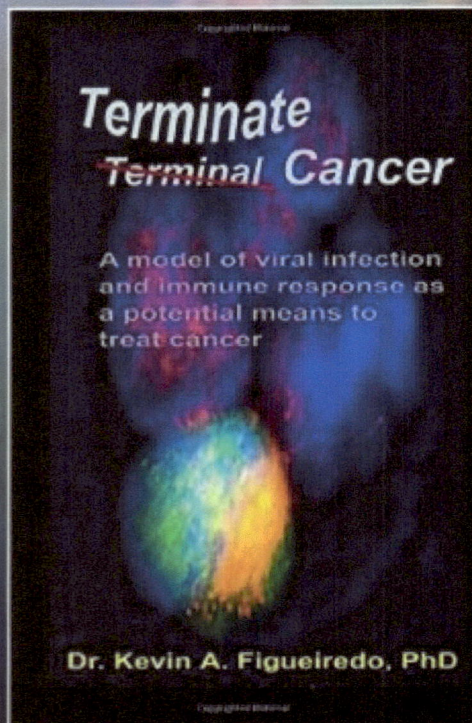

www.ingramcontent.com/pod-product-compliance
Lightning Source LLC
Chambersburg PA
CBHW041213270326
41930CB00001B/8